HOW TO GROW HYDRANGEAS FLOWER

The beginners guide to growing, caring and harvesting hydrangeas at home and garden plus beautiful varieties

Larry Pat

TABLE OF CONTENT

INTRODUCTION

With their enormous flower heads, hydrangeas exude a timeless charm throughout the summer and into fall. These beloved deciduous shrubs, some of which climb, boast striking flower heads in various shapes, ranging from large balls to cones. The mophead and lacecap varieties are the most popular and easily recognizable, displaying large, round flower heads in hues of white, blue, and pink during the summer and autumn months.

Ideal conditions for hydrangeas include moist, well-drained soil and dappled shade – avoiding excessively sunny or shady locations, particularly in south-facing positions with dry soil. Climbing hydrangeas, such as Hydrangea anomala subsp., thrive in heavily shaded spots like north-facing walls.

Hydrangeas cater to diverse preferences with compact options for smaller spaces, double-flowered and bi-colored choices, and varieties that undergo color changes as the flowers mature. Some hydrangeas bloom on both old and new wood, producing flowers twice in a season. The range includes scented varieties and those with captivating autumn foliage. Whether you seek a more traditional style for a cottage garden or shrub border, or a contemporary and urban aesthetic, there's a hydrangea suited to every garden.

About Hydrangeas

Unmatched in the realm of shrubs for their exquisite flowers, these graceful perennial plants are simple to cultivate, adaptable to nearly any soil type, and yield profuse blooms. Their colors captivate with shades of clear blue, vibrant pink, frosty white, lavender, and rose blossoms – occasionally all adorning the same plant!

Hydrangeas prove versatile in various garden settings, whether in group plantings, shrub borders, or containers. The array of available varieties seems limitless, with breeders introducing new options each year, and gardeners' desires for bloom size and color know no bounds. Understanding the specific characteristics of the different species outlined below is crucial for anticipating how your hydrangea will grow, as some may require distinct care. Armed with this knowledge, the joys of your hydrangea's blossoms will be truly enhanced.

PLANTING

The majority of hydrangeas thrive in partial sun, ideally receiving full sun in the morning and partial shade in the afternoon. This is particularly crucial for the Bigleaf hydrangea (H. macrophylla), although some varieties exhibit greater tolerance to full sun. Hydrangeas generally flourish in fertile, well-draining soils that maintain ample moisture. To enhance poor soil, consider incorporating compost or aged manure.

When to plant hydrangeas

For optimal results, the best times to plant hydrangeas are in autumn or spring. Aim for spring planting after the last frost or fall planting before the first frost. This strategic timing allows the shrub ample opportunity to establish a robust root system before the intense heat of summer or the harsh cold of winter, making the cooler shoulder seasons the most favorable planting times.

When planting, opt for the early morning or late afternoon. These times are generally cooler, reducing the likelihood of the plant wilting in extreme heat.

How to Plant Hydrangeas

Space hydrangeas at varying distances, ranging from 3 to 10 feet apart, depending on the type, and always consider the expected size at maturity when determining the spacing.

When planting, carefully remove the hydrangea from its container, checking the root ball for any signs of decay. Trim away any dead or rotting parts and gently loosen the roots if they are tightly bound. Make a very deep hole as the root ball and 2 to 3 times as wide. Ensure that the base of the plant, where the stem meets the soil, is level with the top of the planting hole.

Place the plant in the hole and fill it halfway with soil, followed by generous watering.

Once the water is absorbed, complete filling the hole with soil and water again. This process helps establish the hydrangea in its new location, fostering healthy growth.

How to Grow Hydrangeas from Cuttings

Growing hydrangeas from cuttings is a straightforward process that readily yields roots and provides an excellent lesson in propagation.

Here's a step-by-step guide:

1. Identify a new branch on a well-established hydrangea, specifically one that has not yet flowered and bears three or more pairs of leaves. New growth will appear lighter in color, and the branch will be less rigid.

2. Trim 5 to 6 inches from the tip of the chosen branch, discarding the bottom piece.

3. If the tip cutting has at least two pairs of leaves, remove the lowest pair flush to the stem or node. If the remaining leaves are wide, cut them into two equal halves, removing the tip half. Dust the cutting's end with rooting hormone and, if desired, an antifungal plant powder to promote rooting and prevent rotting.

4. Fill a small pot with moistened potting mix. Plant the cutting. Lightly water to eliminate any air gaps around the stem. Cover the pot loosely with a plastic bag to maintain humidity. If necessary, use chopsticks or kebab sticks to prop up the bag, ensuring it doesn't touch the leaves to prevent rotting. Place the pot in a warm area, sheltered from direct sunlight and wind. Water when the top layer of soil becomes dry.

5. After about a week, gently tug on the cutting. If you encounter resistance, roots have formed. If there's no resistance, check for signs of rotting.

This method is an effective way to propagate hydrangeas and witness the growth process firsthand.

Layering Hydrangeas

For bigleaf hydrangeas during the summer, follow these steps to propagate:

1. Dig a trench alongside the plant, focusing on a branch that extends beyond the trench.

2. Where the chosen branch makes contact with the soil, carefully remove about an inch of the outer bark all around it.

3. Bury the exposed portion of the branch in the trench, securing it with a florist's pin or a lightweight object. Leave 6 to 12 inches of the branch tip uncovered.

4. Water the area regularly to support the rooting process.

5. By early spring, the branch should be ready to be detached from the parent plant and transplanted.

GROWING HYDRANGEAS

Watering Hydrangeas

1. For the initial two years post-planting and in dry spells, ensure hydrangeas receive ample water. If feasible, water in the morning to acclimate the plants to daytime heat and reduce the risk of diseases.

2. Maintain a weekly watering rate of 1 inch throughout the growing season. Deep soaks three times a week, utilizing a soaker hose or a similar method, foster better root growth compared to frequent sprinkling. Consistent moisture is crucial for all varieties, with bigleaf and smooth hydrangeas requiring more water, as insufficient soil moisture can lead to wilted leaves and hindered flowering.

3. Apply organic mulch beneath hydrangeas to retain soil moisture, keep it cool, provide nutrients over time, and enhance soil texture.

Fertilizing Hydrangeas

1. In nutrient-rich soil, hydrangeas usually require minimal fertilizer to avoid excessive leafy growth at the expense of blooms. Conduct a soil test to determine specific fertility needs.

2. Tailor fertilizer application based on the hydrangea variety. For instance, bigleaf hydrangeas benefit from light applications in March, May, and June, while oakleaf and panicle hydrangeas thrive with two applications in April and June. Smooth hydrangeas require fertilization only once, in late winter.

Winter Protection

- During fall, cover hydrangea plants to a depth of at least 18 inches using materials like bark mulch, leaves (excluding maple), pine needles, or straw. If possible, create cages with snow fencing or chicken wire and loosely fill them with leaves to cover the entire plant, tips included.

Avoid wet maple leaves, which may mat and suffocate the plant.

Pruning Hydrangeas

- Pruning techniques vary based on the hydrangea variety, so it's crucial to identify the specific type before deciding on the pruning approach. Understanding the hydrangea variety makes it easier to determine the appropriate pruning technique.

HYDRANGEAS TYPE AND WHEN TO PRUNE

Different hydrangea varieties require specific pruning times and methods. Here's a breakdown:

1. Bigleaf (H. macrophylla)

- Prune after flowering in summer since it blooms on old growth.
- Flower buds form in late summer, so avoid pruning after August 1.
- Only remove dead wood in fall or early spring.
- To encourage branching, cut one or two of the oldest stems down to the base.
- If the plant is old, neglected, or damaged, prune all stems down to the base, sacrificing flowers for that season.

2. Oakleaf (H. quercifolia), Mountain (H. serrata), Climbing (H. anomala ssp. petiolaris)

- Prune after flowering in summer as they bloom on old growth.
- Avoid pruning after August 1 when flower buds form in late summer.
- Remove only dead wood in fall or very early spring.
- For H. macrophylla and H. serrata in Zones 4 and 5, prune only if necessary, immediately after blooming; otherwise, remove dead stems in spring.

3. Panicle (H. paniculata) and Smooth (H. arborescens)

- Prune in late winter before spring growth since they bloom on new growth.
- Prune only dead branches, avoiding shaping the bush.
- Late winter pruning ensures new buds in spring, guaranteeing blooms even if winter buds are damaged.

RECOMMENDED VARIETIES

For a more in-depth exploration of various hydrangea types, refer to "Hydrangea Varieties for Every Garden." There are two primary hydrangea groups:

GROUP 1: PLANTS THAT BLOOM ON NEW GROWTH

Hydrangeas in this category form buds on new growth in early summer, ensuring reliable flowering each year without requiring special care.

Panicle Hydrangeas (Hydrangea paniculata)

- Varieties such as 'Grandiflora' and 'PeeGee' are classic, larger, and more relaxed, while 'Tardiva,' 'White Moth,' and 'Pee Wee' suit smaller gardens. 'Limelight' produces cool-green flowers and reaches a height of 6 to 8 feet.

Smooth Hydrangeas (H. arborescens)
- Look for cultivars like H. arborescens 'Grandiflora' and 'Annabelle,' which yield numerous large, symmetrical blooms, up to 14 inches across, in late summer.

GROUP 2: PLANTS THAT FLOWER ON ESTABLISHED GROWTH

These plants are suitable for those in Zone 8 or warmer. Gardeners in cooler climates may face challenges as these hydrangeas set flower buds in the fall, making them susceptible to damage from early fall or late spring frosts and extremely cold winter temperatures.

- Oakleaf Hydrangeas (H. quercifolia):
- Varieties like 'Snow Queen,' 'SnowFlake,' and 'Alice' promise exceptional fall color.
- Bigleaf Hydrangeas (H. macrophylla):

- All Summer Beauty' (mophead) features profuse dark blue flowers that turn pinker in near-neutral pH soils.
- 'Nikko Blue' (mophead) is robust, displaying large, rounded, blue flowers.
- Blue Wave' (lacecap) produces rich blue to mauve or lilac-blue to pink flowers.
- 'Color Fantasy' (mophead) exhibits reddish or deep purple flowers and grows to about 3 feet tall.
- Mountain Hydrangeas (H. serrata):
- Bluebird' and 'Diadem' are early bloomers, while 'Preziosa' produces extraordinary blossoms with pale shades of blue, mauve, violet, and green in acidic soil.
- Climbing Hydrangeas (H. anomala ssp. Petiolaris):
- Firefly boasts variegated foliage.

HARVESTING

Harvest fully grown hydrangea blossoms in the morning, following a thorough watering of the plant. Immediately place the fresh stems in cold water to prevent wilting. Trim the woody stems underwater at a slant. Eliminate lower leaves from the stems. Assemble the hydrangeas in a vase and position them in a cool spot away from direct sunlight. Monitor and maintain the water level and quality daily, changing it if it becomes cloudy. Revitalize wilting blooms by misting them with water or soaking them in cool water for 10 to 15 minutes.

For Bigleaf Hydrangea (mophead), utilize dried flowers for creative decorations like wreaths. Cut the flower heads once they've matured and acquired a papery texture. Remove leaves from the stems and hang them upside down in a warm, dry, dark, airy room.

Once completely dry (typically within a couple of weeks), store them in a dry location away from direct sunlight. Enhance the flower color by spritzing dry blooms with diluted fabric dye. Explore four alternative methods to dry and preserve your flowers.

Wit and wisdom

The term "hydrangea" finds its roots in the Greek words "hydor," meaning "water," and "angeion," meaning "vessel." This nomenclature alludes to the plant's seed pods, resembling petite water jugs.

In the floral language, hydrangeas convey sentiments of gratitude for being comprehended, or alternatively, they symbolize frigidity and heartlessness.

PESTS/DISEASES

While pests are infrequent, they can emerge when plants undergo stress. To shield your hydrangeas against pests and diseases, opt for resistant cultivars and adhere to our guidelines for proper care.

Potential diseases include Botrytis blight, southern blight, bacterial leaf spot, fungal leaf spot, powdery mildew, Armillaria root rot, Phytophthora crown and root rot, Pythium root and stem rot, Rhizoctonia root and stem rot, rust, and viruses.

Keep an eye out for pests such as aphids, foliar nematodes, root-knot nematodes, stem and bulb nematodes, as well as spider mites. Implementing preventive measures and promptly addressing signs of stress can significantly contribute to maintaining the health and vitality of your hydrangea plants.

Hydrangeas Not Blooming? 5 Causes

If your hydrangea is not blooming, consider these five common reasons:

1. Pruning at the Wrong Time: Be aware of the specific variety of your hydrangea, as some types are pruned before flowering, while others are pruned afterward. Pruning without this knowledge may lead to inadvertently cutting off buds (blooms).

2. Inadequate Moisture: Hydrangeas, as suggested by their name, thrive in consistently moist soil, though it should not be excessively wet. Ensure proper watering practices to maintain optimal moisture levels.

3. Improper Sunlight Exposure: Provide the ideal location for your hydrangea, which includes a few hours of direct sunlight in the morning and dappled sunlight in the afternoon. Too little or too intense sunlight can impact flowering.

4. Fertilizer Choice: If your hydrangea is producing lush green leaves but no blooms, assess your fertilizer. Avoid using nitrogen-heavy fertilizers and opt for ones high in phosphorus (P) during early spring and mid-summer to encourage flowering.

5. Weather and Climate Considerations: Select a hydrangea variety suitable for your climate zone. Even then, unexpected late spring frosts can potentially harm budding plants. If frost is predicted, cover the shrubs with a sheet overnight to protect them.

HOW TO PRUNE

The timing for pruning your hydrangea depends on the type of wood it blooms on. If your hydrangea flowers on "old" wood, pruning in the fall is not recommended, as you risk cutting off next year's flowers.

For both mophead and lacecap hydrangeas, which fall under the bigleaf or macrophylla hydrangea category, pruning can be done right after flowering. This involves cutting back the flowering shoots to the next bud. In cases where older plants are not blooming well, it's possible to cut up to a third of the stems off at the base in late summer. This practice encourages new growth and can rejuvenate the plant.

Understanding the specific pruning requirements for different types of hydrangeas is essential for maintaining healthy and vibrant plants.

Here's a summary of the guidelines for pruning various hydrangea types:

1. Mophead and Bigleaf Hydrangeas (Hydrangea macrophylla)

- Examples: 'Endless Summer' and 'Let's Dance Starlight.'
- These varieties bloom on both old and new wood.
- Pruning is generally not necessary, but if needed, do so immediately after blooming.
- Remove decayed stems only in the spring.

2. Panicle Hydrangeas (Hydrangea paniculata)

- Examples: 'Grandiflora,' 'Pinky Winky,' 'Vanilla Strawberry,' 'Limelight.'
- Panicle types are hardy and suitable for cooler climates.
- Prune in early spring or late winter. While fall pruning is possible,

it's recommended to wait until late winter/early spring to reduce the risk of injury.

3. Smooth Hydrangeas (Hydrangea arborescens)

- Example: 'Annabelle.'
- Native to North America, these hydrangeas tolerate light shade and bloom from June to fall.
- Prune back to the ground in late winter or early spring. Delay fall pruning and wait until late winter/early spring.

4. Oakleaf Hydrangeas (Hydrangea quercifolia)

- Also a North American native with deeply lobed leaves.
- Blooms on old wood.
- Prune just after the flowers have faded and no later.

HYDRANGEA VARIETIES FOR EVERY GARDEN

Hydrangeas are a delightful addition to summer gardens, offering a wide range of types, including the popular Big Leaf and Panicle varieties, distinct Oakleaf, and robust Mountain hydrangeas.

These easy-to-care-for shrubs bloom during the peak of summer, providing weeks of vibrant colors for both gardens and vases. With a diverse array of colors, forms, and sizes, hydrangeas are versatile, suitable for containers, small gardens, mass plantings, and long-flowering hedges.

Bigleaf Hydrangeas (H. macrophylla)

- Commonly known as French hydrangeas, with mop heads or lacecaps.
- Mopheads are hardy in Zone 6, featuring dense pom-pom clusters that change color based on soil pH.

- Lacecaps, suitable for Zones 5 to 9, display intricate flower heads with a combination of tiny fertile buds and showy sterile flowers.
- Recommended varieties include 'All Summer Beauty,' 'Nikko Blue,' 'Blue Wave,' and 'Color Fantasy.'

Panicle Hydrangeas (H. paniculata)

- Named for cone-shaped flower heads, these hydrangeas are easy to grow in various conditions (Zones 3 to 9).
- Blooms on new growth, eliminating the risk of losing flowers to cold winters or improper pruning.
- Varieties like 'Grandiflora,' 'Limelight,' and 'Little Lime' offer beautiful flower clusters in shades of white and pink.

Smooth Hydrangeas (H. arborescens)

- Known as the original "snowball," native to the United States.
- Flourishes in cold climates up to Zone 3, producing large, symmetrical blooms.

- Recommended varieties include 'Grandiflora' and 'Annabelle.'

Oakleaf Hydrangeas (H. quercifolia)

- Native to the Southeast, exhibits remarkable bud hardiness up to Zone 5.
- Features large, coarse leaves resembling those of an oak tree and stunning fall colors.
- Exceptional varieties include 'Snow Queen,' 'SnowFlake,' and 'Alice.'

Mountain Hydrangeas (H. serrata)

- Considered a variety of H. macrophylla by some botanists.
- A smaller lacecap type with leaves showcasing a sawlike margin.
- Hardy in Zones 5 to 9, with 'Bluebird,' 'Diadem,' and 'Preziosa' being notable varieties.

Climbing Hydrangeas (H. anomala ssp. Petiolaris)

- A deciduous vine that blooms from late June to early July.
- Features flat, lacy, creamy-white flowers against glossy leaves.
- 'Firefly' is a newly-patented variety with variegated foliage.

HOW TO CHANGE THE COLOR OF HYDRANGEAS

Altering the color of hydrangea flowers involves changing the soil pH, but it's not an immediate process. It's important to note that not all hydrangea varieties respond to color changes. White flowers, for example, remain unaffected by adjustments in soil pH, as the pH specifically influences the blue and pink tones in certain bigleaf hydrangeas, particularly mophead and lacecap types, as well as certain mountain hydrangea cultivars.

Here's a stepwise guide on how to modify the color of hydrangea flowers:

1. Identify Soil pH

   - Blue flowers are typically produced in acidic soils with a pH below 5.5.
   - Pink flowers are more common in soils with a pH above 6.0.

2. Perform a Soil Test

- Utilize a soil test kit to determine the current pH of the soil where your hydrangeas are planted.

3. Adjusting Soil pH for Color Alteration

- To shift blue flowers to pink: If the soil is excessively acidic (pH below 5.5), introduce lime to elevate the pH.
- To change pink flowers to blue: If the soil is overly alkaline (pH above 6.0), incorporate aluminum sulfate to reduce the pH.

4. Application of Amendments

- Apply lime or aluminum sulfate following package instructions, typically during spring or fall.
- Work these amendments into the soil around the base of the hydrangea plants.

5. Exercise Patience and Monitor

- Changing the color of hydrangea flowers is a gradual process spanning weeks or even months.
- Regularly check the soil pH and make adjustments as required to maintain the desired color.

6. Considerations

- It is advisable for a hydrangea to be at least 2 years old before attempting a pH change to allow it to recover from the initial shock of planting.
- Changing blue flowers to pink is generally considered easier than changing pink flowers to blue.

Which Hydrangeas Can Change Color?

Not all hydrangeas undergo color changes; it's specifically the flowers of certain Bigleaf hydrangeas, especially Mophead and Lacecap types, as well as H. serrata cultivars, that are influenced by soil pH.

Blue hydrangeas thrive in acidic soil with a pH below 5.5, while pink and red varieties prefer alkaline or neutral soil with a pH above 6.0. White hydrangeas remain unaffected by soil pH, and their color, which cannot be changed, typically suits conditions similar to those preferred by pink and red varieties.

The connection between color and pH involves the availability of aluminum ions and a cultivar's ability to absorb them, making the relationship more intricate than a simple numerical pH scale.

How Long Does it Take to Change Color?

Modifying hydrangea colors is a gradual process, taking weeks or even months for the desired changes. Changing blue flowers to pink is generally easier than changing pink flowers to blue, and certain cultivars exhibit more color variability than others.

It is advisable to wait until the plant is at least 2 years old to allow sufficient time for recovery from the shock of the initial planting.

To initiate the color change, conduct a soil test for pH and consult your local nursery for recommended amounts of aluminum sulfate along with specific directions. Once the plant has recovered from its initial planting shock, consider the following guidelines:

Given the information above, adjusting acidity (for blue) or alkalinity (for pink) can be achieved with relative simplicity: [Further details can be provided based on the specific guidelines or instructions.]

From Pink to Blue

To enhance acidity for more intense blue hydrangea flowers, use a solution of 1/4 ounce of aluminum sulfate per gallon of water three times annually. Aluminum sulfate is a clear salt acquired through the reaction of sulfuric acid with hydrated aluminum oxide, readily available at garden centers. Apply this solution to the soil after the plant begins its spring growth, repeating the process at 3- to 4-week intervals. Additionally, once a year in spring, apply 25-5-30 fertilizer following the manufacturer's guidelines.

From Blue to Pink

To raise alkalinity and shift blue hydrangea flowers to pink, apply ground limestone (dolomitic lime) in the spring or fall at a rate of 4 pounds per 100 square feet, ensuring thorough watering. Be cautious about excessive alkalinity, which may result in chlorosis or yellow leaves. Additionally, in the spring or fall, use 25-10-10 fertilizer following the instructions provided by the manufacturer.

Other Factors

As autumn arrives, hydrangea flowers undergo a natural fading and drying process, often displaying a blend of pink, green, or tan hues. This aging process is irreversible.

The color of hydrangea flowers can also be influenced by water quality, especially in regions with hard water containing high mineral content. In such cases, blue flowers may take on a more pinkish tone.

To mitigate this effect, consider using rainwater whenever possible for hydrangea irrigation.

Conclusion

As we reach the final pages of "Hydrangea Haven," the journey you've embarked upon is not just about cultivating flowers; it's about nurturing a connection with nature, creating a sanctuary of beauty right outside your door. The secrets you've uncovered, the techniques you've mastered, and the vibrant blooms that now grace your garden are a testament to the transformative power of understanding and tending to hydrangeas.